THE ERRORS OF PHYSICS

My astrophysical speculations

By Rofer Aspid

Notice to the reader

This book combines widely accepted physical theories with the author's own personal reflections and speculations. Although efforts have been made to present scientific theories accurately, ideas and proposals not supported by empirical evidence should be regarded as mere hypotheses or exploratory thoughts.

The purpose of this work is to stimulate curiosity, critical thinking, and debate, not to offer definitive statements or to replace authoritative academic or scientific sources.

The author assumes no responsibility for misinterpretations or misapplications of the concepts presented. Readers are encouraged to consult additional scientific references if they wish to delve deeper into the topics discussed.

© Rofer Aspid 2024

All rights reserved.

This book, or any part of it, may not be reproduced, stored in a retrieval system, or transmitted in any form or by any means, electronic, mechanical, photocopying, recording, or otherwise, without the prior written permission of the author, except in the case of brief quotations used in critical or scholarly reviews.

INDEX

- PROLOGUE: My astronomical speculations
- On atoms and the nature of light
- On the Theory of Relativity
- The quantum effect or the randomness of movement
- Quarks and the atomic super glue
- Hypothesis on the intrinsic characteristics of the luminous ether
- The singularities of Black Holes
- The electron: a black hole in the microcosm?
- The incongruities of Globular Clusters
- On gravity
- More on gravity
- Theory on light - Bold conclusions
- Why do we feel bad when the "west wind" blows?

PROLOGUE

What has always fascinated me most about particle physics is the realization that, despite the apparent solidity of the objects around me—like the keyboard on which I type or the desk it rests upon—if I could delve deep enough with a powerful microscope, down to the atomic level, I'd discover that matter is astonishingly sparse.

Particles are incredibly light and separated by vast relative distances. What seems so solid to us is, in truth, nearly empty space peppered with tiny particles of matter.

This bears a striking resemblance to astronomy, where stars and galaxies are separated by immense voids of space.

In essence, we live in a world that feels tangible and substantial, but it is almost entirely hollow—a realm of minute particles scattered across immense distances of emptiness. This is true whether we consider the microscopic atomic world or the macroscopic astronomical one.

We find ourselves in a middle ground of scale, observing both extremes, equipped with senses that are practical for daily life but falter when interpreting the very small or the very large. One of my favorite pastimes is to ponder the nature and laws that govern these realms, the particles within atoms and the galaxies across the universe—or perhaps, even universes.

I'm not alone in this; many physicists dedicate their time to grappling with these same questions, striving to reconcile the rules of the quantum world of particles with those of the gravitational cosmos. In other words, they search for a unifying framework that encompasses the four fundamental forces.

I find joy in speculating, indulging in hypotheses that might seem absurd but strike me as beautiful. I like to think of myself as a sentient being made of atoms that didn't originate with me but were forged in a star that exploded long ago and far away—a poetic truth Carl Sagan once taught us.

I also enjoy imagining, though I know it's a bit far-fetched, a striking analogy between an atom and a planetary system. It's still how atoms are often depicted in schools: electrons orbiting a nucleus like planets around a sun. If that's the case, could there be beings composed not of atoms but of planetary systems, where stars serve as nuclei and galaxies as anatomical components?

To us, they would be gods, but who's to say they aren't merely part of an even larger anatomy, built of universes? I like to think of these beings, whose molecules are galaxies, and tell them in modern colloquial terms: "Now *you* are truly immense!"
In the realm of the very large, there seems to be no limit. Even if our universe has boundaries, suggested by the Big Bang, there's the tantalizing possibility of other universes—a "greater" that extends endlessly. In contrast, the very small appears to have limits, at least with our current understanding.

The Planck constant defines a minimum scale below which nothing exists, much like absolute zero in temperature, where molecules cease to move. Just as we can't go below -273.15°C, there is no scale smaller than the Planck length.

The 20th century saw unprecedented advancements in physics, but it was also a time of great mistakes. Esteemed scientists sometimes lost credibility for supporting theories later disproven, occasionally backed by flawed experiments.

One of the most debated errors involved neutrinos appearing to travel faster than light, which would have undermined Einstein's theory of relativity. Immense effort was spent measuring neutrino speeds over hundreds of kilometers, only to find that a clock synchronization error had skewed the results.

Another infamous mistake was the "cold fusion" hypothesis, the idea that nuclear fusion—transforming hydrogen into helium—could occur through chemical means, mimicking the Sun's energy production.

I, too, with a physicist friend, explored alternative interpretations of experiments supporting relativity. For instance, we hypothesized that the deflection of light near a massive object might be explained by a refractive "luminiferous ether," more concentrated near stars, rather than by spacetime curvature due to gravity.

While the ether theory was long discarded, it remains a beautiful concept in my mind. It's tempting to revisit it, especially when modern theories describe space as filled with things we cannot see, such as neutrinos and dark matter. Who knows? Maybe space does harbor something undetectable that explains some of its mysteries—though not neutrinos, which, like invisible men, interact with virtually nothing.

Dark matter, however, which constitutes much of the universe, is thought to accelerate cosmic expansion—an idea that baffled even Einstein. To account for it, he introduced a cosmological constant, though he later regretted it. Still, modern measurements confirm that the universe's expansion is indeed accelerating, undermining the elegant notion of a universe that would one day contract and repeat the cycle.

If this acceleration continues, it dismisses the idea of an oscillating universe or a reversal of entropy—a process as intriguing as it is improbable. Imagine a world where entropy reverses, making everything newer and more ordered, where even time flows backward. Of course, Einstein argued that even in a contracting universe, time would march inexorably forward.

We find ourselves at a crossroads in physics, in dire need of a revolutionary idea to bridge the gap between quantum mechanics and gravity—a unifying insight that transcends our current theories. Perhaps the universe, like us, is a complex structure striving to understand itself.

ON ATOMS AND THE NATURE OF LIGHT

At the beginning of the 20th century, physicists had already understood that when electron beams are directed onto a screen, interference patterns emerge. These can only be explained by considering electrons as waves, which either amplify or cancel each other out according to their frequency, following the precise laws of wave motion.

Davisson discovered that when electrons scatter off the surface of a nickel crystal, they rebound, forming a series of overlapping traces. In 1927, he demonstrated beyond doubt that these overlapping patterns reinforce or cancel each other in a manner consistent with the classical wave interference model. The conclusion was astonishing: electrons behave simultaneously as waves and particles—a concept inconceivable and impossible for us to visualize.

In a different context, according to Newtonian theory, energy is rigorously conserved. It cannot be created or destroyed, only transformed. For instance, an electric stove converts electrical energy into heat, but the total amount remains constant. Because of this fundamental law of physics, perpetual motion machines are impossible—a pursuit that occupied my grandfather's entire life.

However, at the atomic level, this law does not always hold, at least for exceedingly short periods of time due to quantum effects.

Specifically, the conservation of energy dictates that the energy measured at one moment must equal that at the next. But when an electron absorbs a photon, it shifts to a higher energy level. Since an electron can only occupy specific energy levels, if the absorbed energy is insufficient to make the transition, it "borrows" the required energy from nowhere and temporarily reaches the higher state. It quickly emits the photon again and returns to its original state.

The larger the "loan," the shorter the time the electron can remain in the higher energy state. Nevertheless, even this fleeting moment is enough to produce significant effects, such as the emitted photon being redirected—a phenomenon we can observe.

This temporary violation of energy conservation has even more profound implications. For example, matter can spontaneously appear out of nothing, provided it vanishes quickly enough to satisfy the uncertainty principle. In fact, this does happen, and during these brief intervals, spectacular things can be achieved with this "borrowed" energy or matter.

The amounts of "borrowed" energy are minuscule by macroscopic standards, making it impossible to power a machine with such energy. But as a child, I considered my grandfather's idea of perpetual motion absurd. Now that he has passed, I wish I could tell him that his intuition wasn't entirely misguided.

The energy emitted by a light bulb in one second could only be "borrowed," thanks to the uncertainty principle, for a billionth of a billionth of a second. In other words, quantum borrowing amounts to a value so minute—represented by a one followed by thirty-six zeros—that it's negligible on a macroscopic scale. However, in the subatomic realm, where energies are much smaller, even these tiny time intervals allow significant phenomena to occur, such as an electron transitioning between energy levels despite the absorbed photon lacking sufficient energy for excitation.

The photon is re-emitted within a billionth of a second, but in that time, the electron has moved to a different position around the nucleus, and the re-emitted photon will travel in a new direction. This describes how an incoming photon is deflected by the atom.

As the photon's energy approaches what the electron requires for a level transition, the "loan" decreases, allowing the electron to remain longer in the higher energy state, thereby amplifying the dispersive effect.

Since energy is proportional to frequency, which determines the color of light, different colors are dispersed to varying degrees. This explains why some materials are transparent to specific colors and not others, appearing colored when viewed through. The preferential scattering of high-frequency light explains why the sky is blue: sunlight contains many mixed frequencies. High frequencies correspond to blue and violet, while lower ones correspond to green and red. When sunlight strikes air atoms in the upper atmosphere, some blue light scatters, coloring the sky, while the remaining light, rich in lower frequencies, appears yellow. Near the horizon, the greater air depth multiplies this effect, dissipating lower frequencies further, causing the Sun to appear red.

Almost everyone knows that atoms consist mainly of protons and neutrons in the nucleus and electrons orbiting around it. Electrons have a negative charge, protons an equal positive charge, and neutrons no charge.

In recent decades, however, the discovery of numerous subatomic particles has rendered this classification overly simplistic. Thanks to accelerators like cyclotrons, we now know of hundreds of elementary particles. Some have exceedingly brief lifespans, others are so light they rarely interact with matter.

If we collide a particle with an atomic nucleus with sufficient force to disassemble its components, we find that a neutron, once isolated, is unstable. It disintegrates in about fifteen minutes.

Interestingly, when dealing with subatomic particles, the quantum effect makes it impossible to predict when a particle will decay. It could happen instantly or after a long time. Only statistics can reveal when most decay. The term "half-life" describes the time it takes for half of a given quantity of radioactive matter to decay—around fifteen minutes for neutrons.

When a particle decays, the resulting components must obey conservation laws for energy, charge, spin, and other properties. When physicists observed neutron decay into a proton and an electron, they noticed that something was missing. While charge was conserved, the electron's negligible mass left the proton's mass about 1/100 less than the neutron's, violating conservation laws. To resolve this, they hypothesized the neutrino, which remained undetected for 30 years.

The neutrino is extraordinarily light and neutral, neither attracted to electrons nor the nucleus. It can pass through light-years of lead without interacting. Moreover, matter, despite its solidity, is mostly empty space. If we magnified an atom so its nucleus were the size of a soccer ball, the nearest electron would be a grain of rice 10 kilometers away. Thus, what appears solid is mostly emptiness sprinkled with particles.

This explains how neutrinos traverse matter so easily. It also suggests that if matter could be compressed—as it is in neutron stars—a teaspoonful would weigh as much as a mountain.

Contrary to the planetary analogy, atoms are far from mini solar systems. Quantum mechanics reveals a vastly different reality. Neutrinos, despite being elusive and harmless, play a crucial role in the cataclysmic explosions of supernovae.

Today, we produce vast numbers of neutrinos for experimentation. Nuclear reactors, separated from labs by kilometers of mountain to filter other particles, ensure only neutrinos reach the detectors. Even then, interactions with matter are rare. Despite this, scientists have identified three types of neutrinos.

Neutrinos vastly outnumber protons and electrons by a billion to one. The Universe is an immense sea of neutrinos, with atoms as rare impurities. Despite their negligible mass—less than a thousandth that of an electron—their sheer number might outweigh stars and galaxies, dominating cosmic gravity.

Thus, the seemingly inconsequential neutrino may one day hold the key to the Universe's fate, potentially causing its ultimate collapse.

ABOUT THE THEORY OF RELATIVITY

The Theory of Relativity, when first introduced, posed such a challenge to human imagination that only a handful of physicists could grasp it. Even today, it remains so counterintuitive and contrary to common sense that many are tempted to refute it. Undoubtedly, Einstein—who had already earned the Nobel Prize for his groundbreaking work on the photoelectric effect—stands as one of the brightest minds in human history for his remarkable intuition in formulating this theory.

Relativity states that both **SPACE** and **TIME** are not separate entities but rather two aspects of the same fundamental reality. Similarly, it asserts that **MATTER** and **ENERGY** are also two manifestations of the same essence, interchangeable with one another. This equivalence means that matter can be converted into energy and vice versa.

One of the most groundbreaking implications of the theory is that the speed of light is the ultimate speed limit for any particle or object in our universe. The closer an object approaches this speed, the greater its mass becomes, ultimately requiring infinite energy to reach light speed—a feat that is fundamentally impossible.

Years ago, the transformation of energy into matter was confirmed in laboratories where colliding two photons (pure energy) could produce a material particle. On the flip side, we see a dramatic example of matter converting into energy in the devastating power of the atomic bomb—a development that left Einstein deeply troubled due to his indirect role in enabling it. The process is governed by his famous equation, **$E = mc^2$**, where energy equals mass multiplied by the speed of light squared.

The speed of light is crucial in this framework not because of its association with light per se but because it represents a universal constant: the upper limit of speed. Conveniently, light—being pure energy—travels at this speed, which is approximately 300,000 kilometers per second in a vacuum. This same velocity applies to radio waves, neutrinos, and other phenomena.

THE MASS-VELOCITY CONNECTION

The concept that mass increases with velocity is routinely demonstrated in modern particle accelerators, such as cyclotrons and synchrotrons. In these devices, as a particle's speed rises, so does its mass, necessitating increasingly greater amounts of energy to continue accelerating it—until the limits of the accelerator itself are reached.

In everyday life, this phenomenon is imperceptible. Even at the speeds achieved by aircraft or spacecraft, the increase in mass is negligible because these velocities are but a tiny fraction of the speed of light. Importantly, the increase in mass follows an exponential curve: reaching just half the speed of light doesn't mean doubling the mass; far greater velocities are needed for significant changes to occur.

TIME AND VELOCITY

Another profound outcome of Relativity is that **TIME** itself is relative and depends on the observer's velocity. The faster someone moves, the slower time flows for them from the perspective of a stationary observer. This idea is beautifully illustrated in the famous "Twin Paradox."

Imagine one twin embarks on a space voyage aboard a craft traveling near the speed of light. For the traveling twin, only a few months might pass, but upon returning to Earth, they could find that centuries or even millennia have elapsed. In this scenario, the cosmic traveler could hypothetically marry their own great-great-granddaughter, as both would now share the same biological age. However, let it be clear—despite the wild imaginations of science fiction authors—**traveling backward in time is impossible**, as it would require surpassing the speed of light, which our universe forbids.

A HYPOTHETICAL UNIVERSE OF TACHYONS

In another speculative context, one might imagine a parallel universe where all particles travel faster than the speed of light. In such a realm, slowing down to the speed of light would be the challenging barrier. These hypothetical particles, known as **tachyons**, don't contradict Relativity but arise naturally from its equations. If we could interact with tachyons, they might provide a means of instantaneous communication across vast distances —perhaps the method employed by advanced extraterrestrial civilizations. From their perspective, our reliance on radio waves might seem as archaic as smoke signals are to us.

PROOF IN PARTICLE ACCELERATORS

Relativity's predictions about time dilation are verified through experiments involving subatomic particles. Some particles decay mere microseconds after their creation, meaning that even at light speed, they could only travel a few centimeters before disintegrating. However, when these particles are accelerated to near-light speeds, their internal clocks slow down relative to us, allowing them to traverse much greater distances. Engineers and physicists designing particle accelerators must account for this time dilation effect.

THE DUAL LIFE OF A RELATIVISTIC PHYSICIST

A physicist specializing in Relativity lives a paradoxical existence. In everyday life, they share the conventional understanding of time with the rest of us. Yet within their laboratories, they operate under a reality where time is an illusion —a construct without a true past, present, or future.

Many aspects of Relativity and quantum mechanics stretch the limits of human imagination. Yet, the profound questions they address are too significant to remain confined to academic circles. As history shows, physics often leads technological advances by about 50 years. Who knows what breakthroughs await us in the near future, stemming from discoveries made in just the last decade?

ON QUANTUM EFFECTS OR THE RANDOMNESS OF MOTION

Newton stands as one of the most brilliant minds in the history of science. His biography reveals a man endowed with all the qualities of a great scientist, and it is to him that physics owes its foundational momentum to achieve the great advancements of our era. Later, Einstein refined and filled in the gaps Newton could not foresee in his time.

Kepler, building on the ideas of Newton and other contemporaries, established the laws of gravitational motion.

According to our everyday experiences, which also inspired Newton, macroscopic objects set in motion by a force applied in a specific direction will move logically in a straight line unless acted upon by another force.

However, this logic does not hold in the atomic world. For example, a stream of electrons directed from a source toward a specific point does not follow the same path or travel in a straight line. Instead, each electron seems to possess its own "personality," tracing a unique trajectory. Most, indeed, follow the shortest route —the straight line—but others deviate, and the farther the path from the straight line, the fewer electrons choose it. Imagine a group of people crossing a bullring: most would take the direct route, but others might wander off course, with some even stopping to stroll around the arena.

This curious yet real phenomenon is such that no law can describe the exact path any given electron will take. We can only predict statistically what the majority are likely to do. Similarly, it is impossible to predict when a particular radioactive particle will decay—we only know how long it takes for most of them to do so.

But this astonishing randomness goes even further.

Suppose someone claimed that a rubber ball bouncing off a stone wall had a chance—however small—of passing straight through it. Naturally, we would laugh. Yet, such a possibility exists in the atomic world. Electrons directed at an electrostatic barrier —just as impenetrable for them as a stone wall is for the ball— sometimes inexplicably pass through to the other side.

This phenomenon, known as quantum tunneling, is so real that it is exploited in modern electronics. Devices like tunnel diodes depend on this effect. For a rubber ball, composed of countless atoms and electrons, the likelihood of all its particles simultaneously exploiting this tunneling phenomenon is astronomically low but statistically possible.

On another note, how would an electron even "know" the straight line to follow? It doesn't, because its wave-particle duality allows it to explore all possible paths. On these paths, wave interferences cancel out deviations, reinforcing only those closest to the straight trajectory.

Thus, the perfectly ordered motion Newton envisioned is a mere illusion. If the world does not descend into utter chaos at all levels, it is because of an underlying principle that, though less discussed, is no less real: what I would call "cosmic inertia," where nature always seeks the simplest path.

Moreover, if Newton's clockwork universe were entirely true, where every motion was precisely predetermined, our destinies too would be fixed. Though we might remain unaware of the countless laws and initial conditions dictating the cosmos's future, it would undoubtedly be written, leaving us powerless to alter it.

Fortunately, it is this very randomness of motion that grants us freedom—the ability to write our own destinies and influence the cosmos's future.

QUARKS AND THE ATOMIC SUPERGLUE

An electron is a point-like entity surrounded by a swarm of virtual photons and particles that briefly emerge from it before rapidly collapsing back. Among these virtual particles, there are virtual electrons and positrons. Though we cannot observe them directly, we know they exist because they leave measurable physical traces. An electron situated in a vacuum "senses" their presence, as these particles react to its field: virtual positrons are drawn toward the electron, and virtual electrons are repelled, thanks to their respective charges. This interaction leads to a fascinating phenomenon called **vacuum polarization**.

The idea that the vacuum can polarize in the presence of an electric field may seem absurd—it's hard to imagine a vacuum with electrical properties. Yet, this curious prediction of quantum theory has been experimentally validated.

Due to this vacuum polarization, the electron is surrounded by a kind of shield formed by these virtual particles. The electron cannot escape this "cloak" since it is an intrinsic part of its nature. From a distance, the electron appears to have a reduced charge due to this shielding. However, if we could penetrate this layer, we would perceive the electron's actual charge as much stronger.

In our macroscopic world, forces such as gravity and electromagnetism diminish with distance. The farther apart two objects are, the weaker the force between them. Within the microscopic realm of particles, this relationship can reverse. Inside the shielding layer, the farther one moves from the central point-like electron, the stronger the force becomes, as the virtual particles repel with increasing strength.

This phenomenon is pivotal for understanding why, despite the advanced particle accelerators we have today, we have yet to break apart particles like protons and directly observe the quarks they comprise.

THE EMERGENCE OF QUARKS

When experiments with cosmic rays and particle accelerators began, scientists discovered a bewildering variety of particles. These particles often decayed into more stable or long-lived ones. Far from simplifying the understanding of matter's structure, this abundance seemed to complicate it.

It was a situation reminiscent of the periodic table before the atomic structure was understood: seemingly chaotic elements were grouped by their properties, and gaps suggested undiscovered members. Similarly, physicists realized that by proposing the existence of more fundamental particles with specific characteristics, they could explain all known particles. These fundamental entities were named **quarks**.

Initially, only three types of quarks were postulated to account for the particles observed. Later discoveries necessitated the addition of two more quarks, bringing the total to five types. These quarks, in various combinations, form all the heavy particles (muons, baryons, etc.) observed so far.

THE QUARK CONUNDRUM

Despite extensive efforts, no particle has been successfully split into its constituent quarks, nor has any isolated quark been directly observed. This might be due to the shielding effect similar to that of electrons. Within heavy particles, quarks are protected by a barrier of virtual particles. This "screen" causes the force of attraction between quarks to increase as they are pulled apart. Inside the particle, however, the inter-quark forces diminish, allowing quarks to move freely—a phenomenon known as **quark confinement**.

To overcome this virtual particle barrier and isolate quarks would require energies equivalent to those generated by an entire galaxy. As such, it is possible that humanity may never directly observe a quark, constrained by the very nature of the forces that bind them—the so-called **atomic superglue**.

This extraordinary phenomenon underscores the intricate and resilient fabric of matter, governed by principles far beyond the reach of classical intuition, and highlights the boundless mysteries of the quantum world.

HYPOTHESIS ON THE INTRINSIC CHARACTERISTICS OF THE LUMINOUS AETHER

I will start by proposing that light is a wave (and by "light," I mean the entire electromagnetic spectrum). With this assumption, I resolve a significant number of problems and inconsistencies posed by the wave-particle model. Among these, the impossibility of directly visualizing anything akin to this model, and thus having to abandon the use of intelligence in resolving models, limiting ourselves to an act of faith by adhering only to the consequences manifested through the experimentation of phenomena, without maintaining an empirical model that supports the phenomenon.

I will not delve into this topic here, as it would deviate from the proposed subject, but I will refrain from addressing the resolutions of this model for phenomena such as stellar aberration and the photoelectric effect (all other phenomena are easily resolved).

I believe that, in principle, we can state that given the variety of materials capable of supporting the transmission of the luminous phenomenon and electromagnetic waves, where only (due to their inherent characteristics) the propagation speed of these waves is altered, we can assume that these materials themselves are not what comprise the medium that supports the propagation of the waves, yet they substantially modify their characteristics.

Therefore, if we accept that light is a wave like any other, we must also accept the existence of a medium capable of supporting its propagation.

Since the ability to "recover" the space behind the wave in a fluid defines the propagation speed of the wave through that fluid, we must assume that the "ether" (and we shall call it this in honor of our predecessors who defended its existence) is incredibly light or fluid and must "permeate everything."

On the other hand, matter should have the ability to densify this medium around it, yet fluidize it within its core. As a result, inside a body, the propagation speed of light decreases (just as sound travels more slowly through air than through a solid because the particles of air are further apart than those of a solid). Thus, the ether inside a mass is lighter, and light (or electromagnetic waves) will travel more slowly.

That is, within a mass, the "density" of the ether is inversely proportional to the "density" of the matter. However, on the outside, the ether is directly proportional to the density of the matter, meaning that near a mass, the ether becomes denser, causing light to be refracted, which is in turn driven by the gravitational phenomenon. It would also follow that near a mass, the speed of light would be higher than in the "sidereal void."

Verification of this fact, which can be done experimentally, would validate my hypothesis.

The Earth moves around the Sun, which in turn moves around the Galaxy, and the Galaxy moves around a common gravitational center within its local group of galaxies, which are also traveling through space, etc.

Every time we consider the Earth's movement on a larger scale, the speed increases. However, the direction of movement does not affect the speed at which light moves away from us in any direction. The Michelson-Morley experiment demonstrates this, leading to the formulation of the theory of Relativity. Here, I would like to make an aside: to be absolutely certain that this is the case, the Michelson experiment would need to be carried out entirely outside the Earth's atmosphere.

There are instances where even a minimal amount of matter, much less than the Earth's atmosphere, can completely interfere with the phenomenon under study, and this might be the case. The Earth's atmosphere might cause the ether to be completely dragged, making the light behave as though it were entirely dependent on a fluid carried by the Earth.

Likewise, the same should be done with the pi meson experiment in a synchrotron, where, traveling close to the speed of light, it disintegrates into two photons that continue to travel at the speed of light (though having a synchrotron outside the atmosphere is much more difficult today).

Well, let's suppose the experiment is done and we get the same result, i.e., the speed of light is the same regardless of the speed of the medium through which it travels.

This typical behavior of waves would suggest that the ether is completely dragged by the Earth, and through this path, we reach an incongruity, an absurdity. We cannot conceive of a fluid that permeates everything and is simultaneously fully dragged by a moving object.

Questions would arise, leading to absurd answers, such as how much of the ether surrounding the object moves with it? Or what happens when two objects, each with its own ether, cross paths? We are indeed trapped in a dead-end.

In summary, the ether model we propose must have these five main properties:

1. The ether must be incredibly light.
2. It must permeate everything.
3. It must become lighter inside the matter and denser on the outside, with the density increasing the closer it is

to the matter. In interstellar space, it would have an intermediate value.
4. It must be fully dragged by any object in motion and remain fixed.
5. It must have great rigidity to allow the propagation of transverse waves.

A solution to the third and fourth properties is provided by Einstein's relativity. However, to do so, we must eliminate the ether and, consequently, consider light both as a particle and as a wave (impossible to comprehend), while making time and space dilate and contract, etc. We know all the famous relativistic consequences.

Now, let's examine another possibility that was not known at the time when the theory of relativity was formulated. Let's return to our waves and our ether.

For everything to work smoothly, we only need to reconcile the five properties of the "ether." But how can we reconcile such seemingly contradictory facts, like the ether being static, permeating everything, while at the same time being dragged by the medium? Or that it can be light and rigid at the same time? Well, there exists a medium:

Let's suppose that the etheric substance is a property of the universe, whose fundamental characteristic is that its constituent particles appear and disappear in a short span of time, that is, they are made of virtual particles.

In such a substance, whose components are constantly existing and not existing, what happens when an object moves through it? At the moment when the particles do not exist, it cannot be said that the object moves through it, nor that it is stationary. Both things can be said at the same time, if you prefer, for all the qualities stated and many others can be attributed to it.

And when the particles exist, they serve as a support for the wave, ensuring it does not fade. Being something that is constantly born and dies, it simultaneously participates in all the characteristics we thought incompatible.

I believe that demonstrating mathematically these characteristics in both scenarios should not be difficult.

What consequences arise from this hypothesis? In principle, we can now explain the nature of light as a wave without the incomprehensible dualities because we already have our medium for propagation.

Being a quality of the universe, it permeates everything. The speed of light *would* be greater near a mass, thus many relativistic hypotheses could be revoked, and this is a crucial and experimentally demonstrable fact.

We would also need to clarify another fact, such as whether the densification of the ether near a mass is due to a higher concentration of particles or a variation in the rate of their formation due to the presence of that mass.

In the latter case, if time is indeed the rate at which virtual particles form in nature, it could be altered in the presence of the mass, leading to a dilation of time. It would be interesting if the development of this theory led to some relativistic phenomena, though, in this case, we should point out that Einstein, without knowing the causes, arrived at some of the consequences in his famous theory. Now, with the real reasons in hand, we would be in an excellent position to speculate on its effects.

Thus, we could propose a few examples:

Regarding the photon, an essential element in quantum physics, we would no longer need to consider it as a particle, since it could be seen as a single wave. In an ordinary medium, any disturbance in the medium usually forms several waves. If the disturbance is singular, they decrease in amplitude.

But in our model, the independence of a wave caused by a specific phenomenon is entirely plausible, and two or three waves could exist, either with the same or different amplitude, etc. Any composition is possible.

Regarding **gravity**, the subject that most concerns me, I

can propose a hypothesis explainable according to the model I present:

Let us suppose that it is an intrinsic characteristic of the universe for virtual particles to form, and that the quantity or duration of these particles only alters in the presence of mass.

Let us assume these particles tend to form isotropically and, therefore, try to distribute themselves evenly. When there is higher density outside a mass and lower density inside, during the lifespan of these particles, they will tend to move toward the mass, attempting to equalize the density difference.

It's similar to the equalization of pressures between two gaseous media. Well, this "ether wind" will be the cause of one mass being attracted or pushed toward another, ultimately the cause of what we call gravity.

THE SINGULARITIES OF BLACK HOLES

For over seventy years since Einstein formulated his theory of Relativity, it has been subjected to rigorous experimental scrutiny to verify its accuracy, always emerging victorious.

One of the most astonishing relativistic predictions states that time is not absolute, but instead is altered in the presence of gravity. In other words, time passes more slowly in a gravitational field than in its absence, although an observer within that gravitational field wouldn't notice.

For instance, at sea level, time runs slower than at the top of a mountain, since gravity is stronger at sea level. However, this difference is so minute that our senses cannot detect it. Nevertheless, we now have advanced techniques to measure these tiny differences, such as using a MASER on a special spacecraft. The MASER, a refined version of the LASER, uses microwave emissions with very short and incredibly stable frequencies.

By using the MASER cycles as clock oscillators on a spacecraft and comparing them with those of an identical MASER on Earth, calculations show that at an altitude of 10,000 km, time should increase by approximately half of one-billionth of a second when compared to the Earth's surface. In practice, measurements match the theory exactly.

This remarkable effect intensifies as gravity increases. On the surface of a neutron star, the disparity between a clock placed there and one far away could already be one percent.

Stars with greater mass than neutron stars will contract

even further, and their gravity will be even more intense. A star larger than our sun, upon exhausting its nuclear fuel at the end of its life, will collapse to form a body just a few kilometers in diameter.

Unable to support its own weight, it will violently collapse into nothingness in a fraction of a second. Its gravity will be so immense that nothing, not even light, could escape from its interior. A black hole would have formed.

The gravity inside a black hole would be so enormous that time would practically collapse. If gravity were infinite, time would be zero.

A distant observer would deduce that clocks on this surface would be completely stopped, though they wouldn't be able to see the clocks, as no light would escape from the surface. However, an observer falling into the black hole (assuming they survive) would not notice anything unusual in their perception of time; to them, everything would seem the same.

But if they could observe what was happening outside, they would witness in just an instant the birth and death of stars, even the end of the universe, millions upon millions of years for us.

Their perception of time would increasingly diverge from that of the outside world, and when they finally fell into the interior of the black hole, where gravity is infinite, time would stop completely, and they would vanish from our space-time universe.

For an observer outside, watching a spacecraft fall into the black hole, it would appear as if the spacecraft was slowing down as it approached. They would never see it enter; there would come a moment when it seemed to have stopped, and its clocks would have frozen.

The observer inside the black hole, on the other hand, would see the spacecraft outside aging rapidly. In an instant for the person inside the black hole, many years would pass for the person outside.

One consequence of falling into a black hole is that one could never escape from it because this would imply traveling backward in time—leaving before having entered, which is only possible for writers of science fiction.

However, and although it is beyond eternity, the interior of a black hole does not differ too much from our space-time. For example, the passage of time for the observer inside remains perfectly normal, though their fall would be full of hardships. If they fell, say, feet first, the difference in gravity between their head and feet at any given moment would be immense.

Up until this point, all speculations described are consequences of a thoroughly tested theory, but continuing to describe what happens next is certainly more complicated. Indeed, if the object in the experiment falling into a black hole cannot escape it and cannot avoid falling deeper, what is its ultimate fate?

It is speculated that perhaps it might emerge on the other side of the hole, into a different universe, or that it might emerge into our universe but in a different region of space-time, thus making it possible to travel faster than light through space and time. The black hole would be a shortcut, a tunnel to quickly reach distant corners.

We have very little data to confirm or refute these speculations, but the little we do have seems to indicate that this is not the case. The truth, unfortunately for science fiction writers, is that the object would simply cease to exist—that is, it would evaporate from space-time, literally becoming nothing.

Recent research, just a couple of years old, conducted by Stephen Hawking, seems to demonstrate that black holes are not stable and will eventually disappear in a massive explosion of radiation.

This is due to the quantum effect at the edge of the hole, where particles and antiparticles are constantly being formed. Normally, within fractions of a second, these particles recombine and disappear. However, at the edge of a black hole, where gravity is nearly infinite, one particle could be absorbed by the hole, and the other would be pushed away. This would literally "extract" energy from the black hole.

This assertion has profound consequences because, until now, it was believed that nothing could escape from a black hole, by definition. As a result, the black hole would eventually destabilize and explode.

Of course, the black hole formed by the implosion of a star is so large that it would take much more time to die than the age of our universe, so this discovery, if it were only that, would have no practical impact.

What is astonishing is that Hawking believes, and seems to prove, that there are millions and millions of black holes the size of a proton—i.e., tiny ones—whose formation has nothing to do with stellar implosions, but with the Big Bang itself. They formed when densities in small portions of matter were so enormous that they could give rise to these tiny black holes.

According to calculations, our universe is likely filled with them, and in our immediate vicinity, from here to Pluto, there should be two on average.

These small black holes would disintegrate much sooner, so it is possible that this phenomenon is occurring regularly, and we might be able to detect it with our instruments.

The radiation they produce as they disappear could likely be detected from a satellite in orbit. There are plans to dedicate an experiment to verify this phenomenon.

A small black hole of these characteristics would constantly produce about six thousand megawatts, equivalent to six nuclear power plants.

Although this energy would be difficult to harness, because its mass, despite being smaller than a proton, would be greater than that of the Himalayas, and it would pass cleanly through the Earth, so it could not be contained in any container.

Its final explosion would be equivalent to the simultaneous

detonation of TEN BILLION one-megaton atomic bombs, and its only trace would be the enormous emission of high-energy gamma rays, which is what scientists are trying to detect.

If this theory is confirmed, as seems likely, there is another very positive consequence for science: it would finally provide a huge step forward in reconciling two theories that, until now, have worked very well on their own and in their respective fields, but whose unification path was unclear.

I am referring to general relativity and quantum mechanics. This would signify a huge leap forward once we find how gravity is quantized, which would pave the way toward achieving the great dream of physicists: to unify the four fundamental forces: the strong interaction, the weak interaction, electromagnetism, and gravity. When this is achieved, as leading experts predict will happen before the end of the century, we may finally reach the ultimate consequences of theoretical physics, which is to understand the Universe in its entirety.

THE ELECTRON: A BLACK HOLE IN THE MICROCOSMOS?

Almost everyone tends to think of the electron as a tiny ball spinning around the atom's nucleus. However, this idea is far from reality due to the quantum effects that occur both within the atom and in the electron itself, rendering this outdated conception meaningless.

The electron sometimes behaves like a particle and sometimes like a wave. This is known as wave-particle duality.

Additionally, if the electron were merely a tiny ball of matter, it would be easy to understand that when another particle collides with it, the electron would deviate, just as two billiard balls would. But the speed the electron would attain from this collision cannot instantaneously be the same across its entire surface because, for that to happen, the pressure at the point of impact would have to be transmitted at the speed of light throughout the rest of the electron, which, as we know from relativity theory, is impossible.

This suggests that the electron has an internal structure, where the farthest part from the impact takes time to reach the speed of the point of impact, as it takes time for the shock wave to propagate through the electron. Without this assumption, it's inconceivable that the entire electron would move—how could the farthest part from the impact know of the collision at the other end unless it was informed by this pressure wave and set into motion?

But if the electron can compress and squeeze, with a shockwave traveling inside it, it must also be divisible. Therefore, it shouldn't be a fundamental part of matter but should be composed of smaller elements, making it divisible. We wouldn't be dealing with a "lepton," but rather another composite particle, according to this reasoning.

Nevertheless, all the experiments conducted so far with electrons seem to demonstrate that it is indeed a fundamental, indivisible particle of matter.

For us, accustomed to the macroscopic world, visualizing the shape of an electron, which is both a particle and a wave, is so complex that it becomes almost impossible.

The closest visualization might be that it is a point surrounded by an infinite number of virtual photons—photons that emanate from that point, trace a small arc, and then return to that point.

There are infinite possible trajectories and distances, depending on the energy of the virtual photons that emerge from the point. As a result, the energy at that point tends toward infinity.

These virtual photons can, for an incredibly brief moment, form another particle and its corresponding antiparticle, which quickly merge, creating new virtual photons that return to the point.

Thus, visualizing a point surrounded by an infinite possibility of virtual photons, more energetic the closer they are to the point, is, in my view, the closest representation of the reality of the electron.

The electron carries a single unit of negative charge, so when two electrons approach, they repel each other. However, this repulsion does not occur in the way it would seem in the macroscopic world.

For example, if two magnets are thrown toward each other

with like poles facing, they would slow down as they approach until, at some point, they begin to separate, likely following a curved (parabolic) trajectory. As the magnets approach, they exchange photons, which is how each magnet detects the presence of the other.

Each photon exchanged slightly alters the magnet's trajectory, so that, in total, these slight changes give the appearance of a curve, though it is actually a broken line.

Similarly, how does one electron become aware of the proximity of another? When two electrons are hurled toward each other in a straight line (this is a hypothesis, because, as we'll see later, due to quantum effects, we can never be sure of an electron's exact trajectory), at a certain moment, they exchange a virtual photon, with one emitting it and the other absorbing it. At that instant, their directions change sharply, and they move apart.

Naturally, we cannot know which electron emits the photon and which absorbs it, or if both do, or if several photons are involved, in which case there would be multiple sharp changes in direction. This is how one electron detects the presence of another, and it is impossible to pinpoint the exact moment of the trajectory change since the emission and absorption of virtual photons are random.

Returning to the visualization of the electron, note its similarity to a tiny black hole where all mass disappears, from which virtual photons and particles constantly emerge, only to quickly return to the interior.

THE INCONGRUITIES OF GLOBULAR CLUSTERS

How is it possible that spherical objects, composed of thousands or even millions of stars, which do not rotate, are ancient, and are not located in the galactic plane but instead form a massive halo around it, can remain in equilibrium for such an incredibly long time?

Surely, all amateur astronomers have, on more than one occasion, delighted in observing globular clusters, or closed clusters, through a telescope.

Perhaps the most observed of them is Messier 13, located in the constellation Hercules, about 34,000 light-years away and containing around 100,000 stars within a radius of 100 light-years.

With a good pair of binoculars, it can be seen as a diffuse star, but only with a telescope can we resolve the outer stars, while the central ones remain unresolved, even with the most powerful telescopes.

Another famous closed cluster is M-3, in Canes Venatici, on the southern edge of the constellation. It also contains around 100,000 stars within a sphere 65 light-years in diameter, located about 60,000 light-years from us, with a total mass of approximately 245,000 solar masses.

The number of stars in a globular cluster can range from 50,000 to 50 million.

According to these data, if our planet were inside one of these clusters, the nocturnal luminosity from the surrounding stars would be almost the same as what we experience on a full moon night.

Certainly, observing a globular cluster through a good telescope is a beautiful spectacle, but personally, its observation raises a series of doubts I would like to express here.

There are many peculiarities about globular clusters that I do not think are adequately explained, at least they pose a challenge to common sense and Newtonian gravity, and it's not that I trust common sense too much, since Albert Einstein demonstrated that the unusual can sometimes be the truth.

First, what immediately stands out is that most of them are perfectly spherical, which suggests that they have no rotation. If they rotated, the centrifugal force would cause them to flatten, as is the case with galaxies. In fact, some that rotate slightly are also slightly flattened, but these are the minority.

The reason why they are located forming a halo around the galaxy is another peculiarity that we will leave aside for the moment.

They are also extremely old and are primarily made up of red giant stars, although there are also white dwarfs, RR Lyrae variables, long-period Cepheids of the W Virginis type, and some semi-regular stars like R V Tauri. In short, stars not so different from others, which, when the cluster was younger, were younger too.

Perhaps their positions and sizes modified slightly the dynamics and balance of the cluster, but they do not in themselves explain anything abnormal.

Until now, the balance that gives longevity to galaxies, stars, planets around their suns, and so on, has always been composed of multiple forces in equilibrium.

Only in the case of stars has the immense expansive force of the thermonuclear explosions within them been able to counteract the enormous gravitational force and prevent collapse.

When those explosions cease due to fuel exhaustion, the star contracts and then explodes as a nova or supernova, leaving behind the remnant of a white dwarf, neutron star, or even collapsing into a black hole, all depending on its initial mass.

In the case of binary, triple, or multiple stars, we have assumed that they don't collide because they rotate around each other, and the centrifugal force maintains necessary equilibrium.

Some binary stars are so close that the gravitational force between them is so intense that they exchange significant amounts of matter, and that's despite rotating around each other at high speeds.

This has always seemed normal to us, and it is a physical law: the conservation of angular momentum, which states that a higher concentration of mass corresponds to a higher angular velocity. In other words, if two objects move closer, they need to rotate faster to maintain equilibrium.

Everything—Earth, the stars in the galaxy—everything remains in place because gravity is counteracted by the centrifugal force from rotation. Everything except for globular clusters.

Indeed, in a globular cluster, we have a mass of stars, forming a sphere, that doesn't rotate. Yet, it remains in equilibrium from ancient times without the stars crashing into each other.

There seem to be only two forces acting on the stars to maintain this equilibrium: GRAVITY and RADIATION.

Just like what happens in the Sun, the closer a star is to the center of the cluster, the more stars are attracting it toward the center, but at the same time, it receives more radiation from that direction, pushing it in the opposite direction.

This appears to be a simple balance, but I don't see it that way at all—quite the opposite. The phenomenon is far from resembling the equilibrium found in the Sun or in a star during its main sequence.

First of all, if we break down the forces acting on each star in relation to every other star, we find a pair of forces that can be represented as two equal vectors pointing in opposite directions.

According to this, we could cancel out pairs of vectors until only two stars remain, without rotation, at any distance, balanced only by the radiation-gravity pair.

And I ask: without rotation, without centrifugal force, is it believable that one star would keep the other at bay, overcoming its gravitational pull, just by the pressure of its radiation? Personally, I don't believe it.

Now, let's consider a bit deeper. What happens when we move closer to the center of the cluster and analyze the pair of forces acting on a star? Imagine a star with two or three others in front of it, aligned toward the center.

This is more likely the closer we get to the center of the cluster, as the density of stars increases. On one hand, the gravity from these stars will accumulate because gravity seems to act this way—the more mass is in front of an object, the stronger its gravitational pull becomes.

So, if we place the Moon next to New Zealand, we'll feel heavier, the gravity will be greater. The same happens with our example star that has another star behind it. But what about the radiation from the star in front?

It will have to pass through the star in between to reach the one we're analyzing, and to do so, it will face such obstacles that only particles like neutrinos will pass immediately, while more massive particles will be almost entirely absorbed by the star in the middle.

These more massive particles are the ones that would exert the solar wind-like pressure opposing gravity, while the less massive particles (such as neutrinos) would pass through the star and have no interaction with our star.

In short, and based on the above, the radiation-gravity

vector pair will be modified in favor of gravity the closer we get to the center of the globular cluster, thus destabilizing the system and causing it to implode.

But no, that's not how it works. Globular clusters are there, defying our logic. And I ask myself: what's happening? Could there be a force, an unknown and undetected radiation deep within their core that maintains the balance? Or perhaps gravity doesn't act as I've described?

Maybe under certain circumstances, the elusive gravitons behave in a particular way, or they have, if they exist (which I personally doubt), a rather whimsical attitude.

Or perhaps there are forces within clusters that modify their behavior, or maybe the laws of the universe aren't so universal after all and don't apply in clusters... who knows...

When you observe a globular cluster through your telescope, or when you're searching in the dark of night for that eyepiece or Barlow lens to bring out its beauty, think about its peculiarities and contradictions, about its strange equilibrium.

ADDITIONAL NOTE -

The esteemed astrophysicist Ivan R. King is a scholar of globular clusters and the author of an excellent article on the subject published in the prestigious magazine *Scientific American* in August 1985.

In his article, Mr. King discusses, among other things, the extraordinary age of these clusters (around 16 billion years), and places them as the oldest objects in the universe, possibly directly originating from the Big Bang.

What's most curious about the article is the difficulty in explaining their dynamic evolution, which Mr. King resolves through an extraordinary acrobatic move that each star in the cluster supposedly performs, a spiral motion in a spherical plane —something highly implausible.

It seems to me that he entered a mathematical equation into a computer, where the question was what the stars in the cluster must do to produce the observed results. The computer responded with a solution that involved the stars performing some kind of circus acrobatic maneuver.

After carefully studying Mr. King's article, I have come to the only possible logical conclusion: the most profound conviction of the incongruities and secrets posed by the existence of these extraordinary objects known as "globular clusters."

ON GRAVITY

The thorough study of Einstein's theory of relativity, whose validity seems unquestionable today, has only raised more questions for me in the search for a simple explanation—simplicity being the goal every physicist pursues to explain what appears complex.

Einstein recognized the striking similarity between gravitational force and inertial force, to the point of considering them as essentially the same.

Indeed, who has not felt the force exerted on them by the backrest of a car seat when it accelerates? Moreover, inertial force can perfectly counterbalance gravitational force. Imagine we are aboard a spacecraft, far from any planet, moving in a straight and uniform motion.

If we attempt to deviate from our trajectory using the ship's engines, we will feel as though we are being pulled toward the opposite wall of the ship. If we were to follow a circular path, we would be pushed outward from the circle. Thus, without looking outside, we can easily deduce the type of movement we are undergoing.

Now, imagine that while sleeping, we unknowingly enter the orbit of a planet. Upon waking, if we do not look out the window, we would think we are still moving in a straight line. However, upon glancing outside, we would realize we are orbiting the planet. No instrument, no experiment, could detect the difference.

This example is crucial for two reasons. First, it clearly indicates the similarity between GRAVITY and INERTIA (to the point that they are the same). But most importantly—and this is a lesson we should take to heart—it teaches us not to rely solely on our senses or logic.

The world of physics and science is full of paradoxes that contradict what we call "common sense," and they are true, having been difficult to uncover because our intuition is guided by everyday experience, which easily deceives us in the face of seemingly implausible situations.

In our daily experience, however, we do not perceive inertia and gravity in exactly the same way. This is mainly because inertia manifests as a force opposing any change in our state of rest or motion, and can act in any direction, while gravity always attracts us toward the center of masses. Yet, envisioning a compatibility is not impossible.

Consider, for example, what would happen if everything we know were constantly increasing in size. Just as the universe expands, as discovered by Hubble, if everything—the atoms, particles, light, space, and time—were continuously growing, we would have no way of noticing because any "meter" we used for measurement would grow in proportion.

Moreover, wouldn't there exist an inertial force manifesting exactly as gravity does?

According to Einstein, gravity is a distortion in spacetime, a curvature caused by mass in our four-dimensional universe.

There are four fundamental forces in nature:
- **GRAVITY**
- **STRONG FORCE**
- **ELECTROMAGNETISM**
- **WEAK FORCE**

Einstein's lifelong dream was to unify these four forces into a single common cause that would explain them all. It remains the dream of almost all physicists. Achieving this would constitute the greatest scientific accomplishment in human history.

Gravity corresponds to the attractive force that objects experience due to their mass, and it can act over vast distances.

Between 1900 and 1930, two famous physicists presented theories that have led to the greatest intellectual, cultural, and technological successes in human history. I refer to Einstein's theory of Special Relativity and Max Planck's quantum theory, which were later expanded by Einstein with his General Theory of Relativity and by other physicists in the case of quantum theory.

To date, both theories are fully confirmed and have led to the development of modern atomic physics.

The strong force is responsible for holding protons together within the atomic nucleus. Without this force, they would separate, as they all carry a positive charge. It only acts at very short distances.

Electromagnetism is the force of attraction or repulsion due to electric charge or magnetism. It operates like gravity over large distances.

The weak force is responsible for atomic radioactivity.

Electromagnetism, in turn, unites two seemingly distinct phenomena into a single one, as the same cause motivates both. Its discovery was a major breakthrough in physics and a stimulus for attempting the same unification with the other "forces."

For one object to influence another, or receive influence from it, it must somehow communicate its presence. This, it is believed, is the only way "forces" can act. Additionally, this information cannot travel faster than the speed of light, which is a condition of the theory of relativity.

The great achievement in unifying electromagnetism was discovering that this force corresponds to the exchange of photons between objects under study, both in the case of electricity and magnetism.

Encouraged by this success, physicists sought to find other particles responsible for the other forces. Recently, this has been achieved in the cases of the strong and weak interactions. It is now hypothesized that gravity must also involve a particle, which is called the graviton.

The characteristics of this particle are presumed to be similar to that of a photon: massless, but with a spin of two rather than one, like the photon. However, to date, no experiment has detected a graviton, which is concerning because if they exist, they should interact very easily with matter (unlike neutrinos, which are so light they can pass through light-years of lead without colliding, yet have been detected).

The universe our senses detect has three spatial dimensions and a fourth dimension—time. Einstein's great intuition was to relate space and time in such a way that, contrary to our common sense, they are one and the same; only the observer's speed transforms one into the other.

This condition dictates that the presence of mass creates a curvature in our four-dimensional spacetime, just as placing a weight on an elastic sheet (two dimensions) would cause a ball placed elsewhere on the sheet to roll toward it.

MORE ABOUT GRAVITY

Throughout my life, from my earliest childhood, I have felt a deep concern—almost obsessive—about understanding what it is and what causes that everyday phenomenon known as the force of gravity.

When I first read the Theory of Relativity, I was so impressed by its illogical principles, so contrary to common sense yet so true and continually demonstrable through the most meticulous scientific experiments, that my teenage dream was to become a Nuclear Physicist, and my hero was Albert Einstein.

Over the years, I have never stopped reading everything I could find about the life and work of my hero. I have spent many hours quietly pondering this mysterious force of GRAVITY, always arriving at the same conclusions, sensing that I am somehow close to the truth.

With time, and without ever losing my respect for the great scientist and human being that Mr. Einstein was, I have come to consider that perhaps in some areas he was not entirely correct. His discoveries were so respected that only a few scientists have since dared to challenge certain aspects of them.

Even Albert himself acknowledged his mistakes on occasion, and in other instances, time has shown that theories he was initially reluctant to accept—such as quantum theory, which he famously rejected with "God does not play dice"—must now be embraced as one of the cornerstones of modern physics.

When I studied the principles behind electromagnetic

attraction and the strong force, I, like everyone else, thought that gravity would be a similar phenomenon.

And, like everyone else (those who care about such things, of course), I read as much as I could about particles, until my brain became overwhelmed by the plethora of them discovered.

I speculated on the quarks, initially thinking that three would be enough to explain everything, then wondering if four would do, eagerly awaiting that "periodic table" of particles and developing a particular fondness for certain particles like neutrinos, which had certainly intriguing characteristics.

All of this while hoping that the elusive graviton would appear one day, despite the fact that its physical characteristics—spin, etc.—are theoretically known. I dreamed of the long-sought UNIFICATION of the four great forces and immersed myself in the study of the Big Bang, pondering what must have happened in those first, definitive moments (or seconds) that shaped the laws of physics, laws whose universality I considered to be beyond question.

I reviewed the unidirectionality of time and became fascinated with the obvious, yet crucial, concept of increasing ENTROPY. Yet, through all of this, I have always had and still hold the conviction that the phenomenon of GRAVITATION is not due to the exchange of particles; that searching for the graviton is futile, because it does not exist—just as once there was no "ether," or the speed of light did not depend on the movement of the emitter, no matter how implausible it seemed, and despite Michelson and Morley repeating their experiment (which, however, rightly earned Michelson the first American Nobel Prize and inspired Einstein's Theory of Relativity).

It is also impossible, using the methods currently employed, to find a way to unify the four great forces, which has been the life's work of many great contemporary physicists, yet without success.

I believe that what happened is that, moments before the Big Bang, the virtual lump accelerated its growth, and space, time, and the universe grew—or, better put, the foundations for what would later become these concepts expanded—while entropy decreased.

But after a few moments, when entropy reached its minimum, the lump became unstable, leading to the Big Bang, starting a DECELERATION. Though everything continued to grow, it did so NOT ACCELERATING, but rather DECELERATING, as entropy began to increase, accelerating more and more.

And this is the crux of the matter. It was like a soap bubble that inflates to a limit, perfect and spherical, then bursts, and its contents scatter.

The reason it happened this way, and not otherwise, can be explained in many ways, some quite obvious. "Why else would it have been any different?" What we know so far does not exclude this possibility.

Just as we say that the universe is the way it is because, had it been otherwise, it would not have given rise to the laws that govern us, and life as we know it might not have been possible, so we wouldn't be here debating this. However, this does not exclude the existence of other universes formed by other bubbles, with laws identical, similar, or different from ours, but, of course, with no possibility of interacting with our own.

If this is true, and if the theory could be demonstrated to be correct, in line with all observable physical laws, then the SPEED OF LIGHT would not be CONSTANT in itself, but subject to CONTINUOUS DECELERATION.

Of course, since space would also shrink (here I refer to the Theory of Relativity regarding space and time), it would be impossible to measure this variation from within our Universe-Time context. However, if the speed of light does decrease for an observer situated outside our space-time context, the space (interatomic space, interparticle space, etc.) must DECREASE proportionally, in such a way that what manifests itself as GRAVITY is merely the CONSTANT VARIATION IN INERTIAL VELOCITY, caused by this interatomic reduction (confirming much of the speculation about the similarities between inertia and gravity).

I agree with the well-known statement that mass deforms SPACE-TIME, curving it, and I assume that not only mass but also energy, which in essence is the same, contributes to this deformation.

That is, proportionally, and considering the mass-energy conversion formula ($E=MC^2$), an astronomical body attracts a photon emission just as the beam of light attracts the mass of the body. This is often overlooked when thinking about the subject, yet it is crucial for drawing conclusions. However, I believe this deformation is a CONSEQUENCE, not a PRINCIPLE, much like trying to explain the inexplicable with words.

The concentration of mass or energy in space produces a gravitational effect on another mass or energy, directly proportional to its quantity and inversely proportional to the square of the distances between them (as Newton already discovered), due to the concentration of the INERTIAL PUSH mentioned earlier.

One consequence (and here lies a challenge for theoretical experimenters to demonstrate the theory) is that if a mass were to disappear instantaneously, its effect would cease immediately as well. (If the Sun were to suddenly disappear, we would continue receiving its light for eight more minutes, but our trajectory would INSTANTLY change to a tangential path to the current one. Of course, this is a figure of speech—I know that we are following a straight path around the Sun in relativistic SPACE-TIME).

I sincerely believe that if the speed of light were constant, gravity as we understand it would not be possible, since I regard it as an inertial manifestation of the negative increase in the speed of light.

Following this line of reasoning, it can be explained, as one might expect, and through means different from those used by Stephen Hawking, why black holes cannot be so black. They must emit radiation, and their life must be limited as well.

It is possible, based on this theory, to go further. With the right mathematical knowledge, we could calculate, based on the relationship between mass, energy, and the speed of light, the negative increment in the speed of light required to satisfy the gravitational constant.

Afterward, knowing the age of the universe, we could calculate its total mass, finally resolving the question of whether the expansion will continue indefinitely or if it will eventually reverse, as many of us hope, at some point, starting again to decrease ENTROPY and changing the direction of TIME.

THEORY OF LIGHT - BOLD CONCLUSIONS

From now on, when we refer to LIGHT, we mean any electromagnetic manifestation, ranging from radio waves (the lowest frequencies) to ultraviolet light and cosmic rays.

PRINCIPLES:

1. LIGHT is *exclusively* a wave-like motion, and as such, it requires a transmission medium.
2. LIGHT utilizes *quantum space*, that is, the constant formation and annihilation of virtual particles, as its transmission medium.
3. TIME is the rhythm of the formation and cancellation of virtual particles.

CONCLUSIONS:

1. Matter interferes with the formation of this transmission medium, which becomes "lighter" or less dense within, thereby explaining the slower speed of LIGHT in a medium as its material density increases.
2. Matter, depending on frequencies, atomic arrangement, etc., either favors or cancels the propagation of waves through it.
3. The fact that the transmission medium is made up of virtual particles explains why no measurable translation of the Earth occurs through the medium, because the medium always moves with us (Michelson-Morley experiment).

4. Time shortens as we move through it because, for an external observer, a traveler encounters more virtual particles, which, however, remain the same for the traveler, up to a limit defined by the very rhythm of particle appearance. (A new explanation for the relativistic phenomenon).
5. The increase in mass of a particle as its speed increases is due to encountering a greater number of virtual particles in its path. What increases is not the mass, but the inertial force of that mass. (A new explanation for the relativistic phenomenon).
6. Inside a black hole, mass leaves no room for the formation of the transmission medium for light. Only virtual particles exist at the event horizon, which is why light cannot be transmitted or propagated inside a black hole. Thus, it is not gravity that causes the opacity of black holes, but rather the absence of the transmission medium.
7. Gravity, as a force, is the manifestation of the change that locally occurs in the formation of virtual particles. As it decreases, this manifests as an increase in mass in relation to the surrounding space. Hence, gravity always manifests as an attractive inertial force. (This point requires a detailed and subsequent development).
8. The bending of LIGHT as it passes near a gravitational field, as well as the negative time dilation in the presence of gravity, can be explained by the variation in the formation of virtual particles in the gravitational environment. Inside a black hole, where no virtual particles are formed, time is not zero; it simply does not exist as a concept.
9. If time is defined by the rhythm of virtual particle formation, which in turn determines the speed of LIGHT, it would be possible, knowing these parameters,

to uncover others, such as the relationship between density, size of these virtual particles, and their average lifetime.

10. Relativity is a consequence of the above, not the primary cause, which I am presenting. It should be noted that Einstein developed it before Planck developed quantum theory. This distinction is fundamental, as it could change the concept of the Universe and possibly lead, through this avenue, to the long-sought UNIFICATION OF FORCES.

Note. - To conclude this book and given that the phenomenon I am going to describe is related to particle physics, I am going to explain why we feel bad when in some coastal areas, such as Valencia, the west wind blows:

Why Do We Feel Bad When the "Poniente" Wind Blows?

For all of us living on the Mediterranean coast, the sensation of discomfort when a type of wind called "poniente" blows is well-known. It comes from the west and is hot and dry, in contrast to the humid, cool wind from the east, called "levante." This feeling of discomfort is also felt in other parts of the world, with winds from other directions and names.

We typically attribute this discomfort to the heat and dryness of the air, but this is a mistake; it is only minimally due to these factors. In fact, in many places, the climate is dry and warm without this discomfort being present.

The cause lies in a phenomenon related to *atmospheric ionization*.

As we know, the air is made up of molecules, which in turn are formed by atoms. Atoms usually carry a neutral electrical charge, where the positive charge of the nucleus is exactly balanced by the negative charge of the electrons in the outer shell. However, this balance is often disturbed by various factors, causing the atom to either lose or gain electrons. In the former case, a positive ion is formed, and in the latter, a negative ion.

This atmospheric ionization is the cause of phenomena like storms. When different layers of air rub against each other, they either lose or gain electrons and become electrically charged. In the pursuit of natural balance, these discharges occur as lightning.

It has been scientifically proven that the concentration of ions, either positive or negative, plays a key role in how we feel. Negative ions make us feel good—relaxed and calm—while positive ions are associated with discomfort, unease, and even depression if absorbed continuously.

The "poniente" wind reaches our coast after traveling many kilometers near the ground from the Castilian plateau and is air laden with positive ions. This is the primary cause of our unease. Many devices in our industrialized civilization, such as televisions, computers, fluorescent lamps, motors, and other machinery, also produce positive ions. This is why working in enclosed spaces surrounded by machinery can lead to stress and depression over time.

Air is full of harmful positive ions before a storm (causing a feeling of oppression), while beneficial negative ions are produced after the storm or after heavy rain. A large concentration of negative ions is also found at the seashore, under dense vegetation, or near a waterfall, as water falling generates negative ions.

There are devices, called ionizers, that specifically produce these beneficial negative ions. They work based on the so-called *corona effect*, which occurs when a very high voltage is applied to a pointed surface.

Ionizers create a sense of well-being, which is why they are increasingly used in rooms where we spend a lot of time. They also have other interesting effects, such as eliminating environmental dust (by binding it to walls), which is especially helpful for people with asthma or allergies to dust and pollen.

It's fascinating to see how placing an ionizer in a container full of cigarette smoke causes the smoke to disappear in just a few seconds, making ionizers useful in smoke-filled environments, such as cars used by smokers, etc.

Although ionizers are quite simple to build (requiring very few electronic components), they operate with very high voltages, so they can be dangerous. Therefore, they should always be purchased properly certified and should never be opened. Otherwise, they are highly recommended.

THE END

www.ingramcontent.com/pod-product-compliance
Lightning Source LLC
Chambersburg PA
CBHW070416230526
45471CB00006B/2837